Navigating the Dangers of Flesh-Eating Bacteria

A Comprehensive Guide to Identification, Treatment, and Prevention of Vibrio Infections

Susan T. Brown

All rights to this publication are reserved. No part of this work may be reproduced, distributed, or transmitted by any means—whether by photocopying, recording, or other electronic or mechanical methods—without prior written permission from the publisher. The only exceptions are brief quotations used in critical reviews or permitted noncommercial uses under copyright law.

Copyright 2024, Susan T. Brown.

Table of Contents

Introduction

Understanding Vibrio Vulnificus

The Impact of Hurricanes and Floods

Identification of Vibrio Infections

Diagnosing Vibrio Infection

Treatment Options

Prevention Strategies

The Role of Public Health

What to Do After a Flood

Personal Stories and Case Studies

Resources and Additional Reading

Introduction

Vibrio vulnificus is a harmful bacterium that lives in warm brackish waters. It flourishes in coastal settings, particularly in warm weather, and becomes considerably more common following hurricanes and tropical storms.

The bacteria can infect humans in two ways: through open wounds that come into touch with polluted water, or by eating raw or undercooked shellfish, such as oysters. While most bacteria are generally harmless, Vibrio vulnificus is aggressive and can cause serious illnesses such as necrotizing fasciitis (also known as flesh-eating disease), sepsis, and even death in some circumstances.

Its capacity to induce immediate tissue damage and blood infections makes it one of the most hazardous bacteria found in warm coastal areas. What distinguishes this bacterium is not just its rapid spread, but also its lethality. Infections can be lethal within days, and survivors are frequently forced to undergo severe measures such as limb amputation to prevent the infection from spreading further.

The Importance of Awareness and Education

Awareness is the first step toward prevention, especially with a hazard as severe as Vibrio vulnificus. Many people who live in coastal areas or visit beaches are unaware of the dangers that warm floodwaters provide.

Without awareness, people may swim or walk in contaminated waterways, unaware that a minor cut on their skin could expose them to a life-threatening infection. Similarly, eating inadequately prepared shellfish without understanding the risks might result in serious gastrointestinal illness or worse.

Everyone should be educated about Vibrio infections, not just those at high risk (such as those with weaker immune systems or chronic liver illness). The microorganism is not discriminatory. It affects both healthy and vulnerable people. However, persons with underlying health conditions are more likely to have serious consequences or die as a result of infections.

Communities affected by disasters must know how to defend themselves. Following a storm, floodwaters can mix sewage, debris, and germs such as Vibrio vulnificus with the water we come into contact with, whether while swimming or cleaning up. People who have open wounds, cuts, or abrasions should avoid these waters or use waterproof bandages to protect themselves.

One of the reasons Vibrio infections remain a public health concern is a lack of general understanding. Often, people are unaware of the bacteria's potential for harm until it is too late. By increasing public awareness, we can significantly reduce the incidence of infections, fatalities, and serious health effects.

For example, during and after hurricanes, public health announcements warn about the dangers of flooding, but they frequently do not elaborate on the specific concerns posed by Vibrio vulnificus. This handbook seeks to close that gap by giving clear and practical information.

In addition to creating awareness among individuals, this approach emphasizes the importance of community-wide education. Beaches, restaurants, and public health authorities all play important roles in reducing the risk of Vibrio infections. We can create a safer environment by working together to educate people about the dangers of contaminated water and seafood, particularly in the aftermath of natural catastrophes such as hurricanes.

In short, knowing more about Vibrio vulnificus allows us to better safeguard ourselves and our communities. This guide is intended to be a thorough resource, including not only data but also actionable measures to keep safe and avoid life-threatening situations.

Vibrio vulnificus may not be a familiar word, but it poses a severe threat in coastal waters, particularly during and after hurricanes.

Because of its propensity to induce quick and serious infections, everyone should be aware of the dangers. Increased information and prevention can help to minimize the number of infections, save lives, and limit the catastrophic impacts of this deadly bacterium. By reading this guide, you are taking a crucial step toward remaining educated and safe.

Understanding Vibrio Vulnificus

Vibrio vulnificus is a bacterium that lives naturally in coastal and brackish environments, particularly where salt and fresh water mingle. It is not a newly discovered organism, but it has received substantial attention in recent years due to the serious illnesses it can cause in humans. This bacterium is a member of the Vibrio family, which includes other hazardous species, but Vibrio vulnificus is notable for its aggressive character and capacity to cause life-threatening infections.

Unlike many bacteria that cause common illnesses, Vibrio vulnificus can cause major health problems quickly. The bacterium can enter the body through open wounds, scrapes, or abrasions exposed to contaminated water, or by consuming raw or undercooked seafood, particularly shellfish such as oysters. The bacteria thrive in warm temperatures, thus it is more common in the summer and areas with warm seas all year.

When Vibrio vulnificus enters the body, it can cause infections that progress swiftly, resulting in serious tissue damage, blood infections, or even death if not treated properly. Understanding what Vibrio vulnificus is and how it behaves is critical because it allows us to protect ourselves from its destructive consequences.

Natural Habitats and Conditions Favorable to Growth

Vibrio vulnificus prefers warm, coastal waters. It favors areas where saltwater and freshwater mingle, such as estuaries, bays, and rivers along the coast. The bacterium is most abundant in waters with intermediate salinity and mild temperatures, which explains why it is more active during the warmer months. Vibrio infections are most common in the summer and early fall, when temperatures range from 68°F to 95°F (20°C to 35°C).

Warmth, as well as natural environmental changes such as storms and significant flooding, promote rapid bacterial growth. Following such storms, floodwaters combine with sewage, pollution, and other toxins, forming a breeding habitat for Vibrio bacteria.

As the waters recede or become stuck in stagnant pools, the risk of human exposure rises. Coastal locations, particularly those hit by storms, are often hotspots for Vibrio outbreaks because the bacteria flourish in the disturbed, nutrient-rich water left behind.

People who live or visit locations with warm, brackish waters are most vulnerable, especially if they come into contact with the waters while wearing open wounds or if they ingest seafood that has been exposed to contaminated waters. Fishermen, beachgoers, and those who engage in water-based activities such as boating or wading in these areas should exercise extra caution during peak Vibrio seasons.

Infections Caused by Vibrio vulnificus

Vibrio vulnificus can cause three types of illnesses: wound infections, septicemia (blood infection), and gastrointestinal disease after eating contaminated seafood.

1. <u>Wound Infections:</u>

Vibrio vulnificus infections are most well-known for entering the body through an open wound. This could be a tiny cut, scrape, or even a puncture caused by a shell or sea debris. The infection can spread quickly, causing the tissue around the wound to degrade and die in a condition known as necrotizing fasciitis, sometimes known as "flesh-eating disease." If not treated promptly, this infection can lead to amputation. In severe situations, the infection can result in sepsis, a life-threatening illness in which the body's response to infection causes widespread inflammation, ultimately leading to organ failure and death.

2. Septicemia (blood infections):
Vibrio vulnificus can enter the bloodstream through an infected wound or the digestive tract after consuming contaminated seafood. Bloodstream infections are very harmful because they spread fast throughout the body, resulting in symptoms including fever, chills, low blood pressure, and blistering skin sores. Without rapid medical attention, septicemia can progress to septic shock, organ failure, and death.

3. Gastrointestinal illness:
Vibrio vulnificus gastrointestinal infections are most commonly caused by the

consumption of raw or undercooked shellfish, particularly oysters. Symptoms typically appear within 24 hours of consuming contaminated food and include vomiting, diarrhea, abdominal pain, and fever.

While many gastrointestinal infections are minor, those with underlying health problems like liver illness or compromised immune systems are far more likely to have serious complications, including blood infections.

Key Statistics for Infection Rates and Mortality

Vibrio vulnificus may not infect as many individuals as other bacterial illnesses, but the consequences are deadly. The infection rate is relatively low, with just 150 to 200 cases recorded in the United States each year, according to the Centers for Disease Control and Prevention (CDC). The fact that this bacterium has such a high mortality rate is extremely troubling.

About one in every five people infected with Vibrio vulnificus will die as a result of the infection, which can occur one or two days after symptoms appear, especially if the infection spreads into the bloodstream or necrotizing fasciitis develops.

In Florida, where warm seas and hurricane-related floods are typical, infection rates have increased after natural catastrophes. For example, following Hurricane Ian in 2022, there were 74 cases and 13 deaths associated with Vibrio vulnificus infections. These figures are not unique to Florida; other coastal locations have observed similar increases in infections after hurricanes or tropical storms.

People with weakened immune systems, especially those with chronic liver illness, are particularly vulnerable. In fact, over half of all fatal instances involve people who have underlying illnesses. This susceptibility emphasizes the necessity of both personal awareness and public health actions, especially in locations prone to natural disasters that provide ideal conditions for Vibrio growth.

Vibrio vulnificus is a potent and hazardous bacterium that thrives in certain settings and situations. Its capacity to cause severe infections, including life-threatening wounds and blood infections, makes it a public health risk, particularly in coastal locations during the warmer months or after hurricanes.

The Impact of Hurricanes and Floods

Natural catastrophes, particularly storms and major flooding, can have serious consequences for public health, notably bacterial illnesses caused by Vibrio vulnificus. While Vibrio bacteria are naturally found in warm coastal waters, cyclones and flooding significantly enhance the likelihood of human exposure to this dangerous bacterium.

How Natural Disasters Affect Vibrio Infections

Hurricanes and floods are especially dangerous in terms of spreading Vibrio bacteria for a variety of reasons. First, storms create widespread flooding, resulting in the mixing of saltwater from the ocean with freshwater from rivers, lakes, and storm runoff. This brackish water provides the ideal environment for Vibrio bacteria to thrive. Following a hurricane, flooded areas contain not just larger amounts of these germs, but also additional toxins such as sewage and industrial waste, contributing to a highly contaminated and deadly environment.

Warm temperatures associated with hurricanes and tropical storms, particularly in Florida and throughout the Gulf Coast, stimulate the growth of Vibrio bacteria. Warm water serves as a breeding habitat for germs, helping them to grow fast.

In the aftermath of a hurricane, stagnant water left behind in flooded areas provides more growth possibilities for Vibrio, increasing the possibility of human contact. Even after the floodwaters have receded, germs can survive in puddles, brackish ponds, and any remaining water sources.

During hurricanes and floods, humans are frequently exposed to Vibrio vulnificus in two ways: by direct contact with contaminated water and by consuming seafood that has been exposed to polluted waters. People with open wounds, scrapes, or even minor cuts are at a high risk of infection if they wade through floodwaters or come into touch with contaminated water. Individuals who consume raw or undercooked seafood, particularly shellfish such as oysters harvested from contaminated locations, are also susceptible to gastrointestinal illnesses or, in severe circumstances, blood infections.

Case Studies: Hurricanes Ian, Helene

The link between hurricanes and increased Vibrio infections has been widely documented in recent years, with Hurricanes Ian and Helene serving as vivid illustrations of how natural catastrophes can cause surges in infection rates.

Hurricane Ian, 2022:

When Hurricane Ian made landfall in Florida in 2022, it brought not just enormous destruction but also an increase in Vibrio vulnificus infections. The storm's massive flooding resulted in extensive mixing of saltwater and freshwater, providing ideal circumstances for Vibrio bacteria to proliferate.

Following the hurricane, Florida's health officials recorded a record number of Vibrio illnesses. Before the hurricane, the state had seen a typical number of cases, but after the storm, the numbers increased dramatically, with 29 new infections reported in the weeks following Ian's landfall, bringing the total number of cases that year to 74. Tragically, these illnesses killed 13 people.

Hurricane Ian emphasized the dangers that occur after natural catastrophes, as individuals waded through flooding during cleanup activities, frequently ignorant of the increased risk of infection. Many of the victims had pre-existing medical issues that made them more vulnerable to the bacteria's rapid spread after they were exposed.

<u>Hurricane Helene, 2024:</u>
A similar event unfolded in 2024 with Hurricane Helene. The hurricane hit Florida at Category 4 strength, pouring tremendous amounts of rain and forcing storm surges far inland, resulting in severe flooding. Before Helene, Pinellas County has recorded no cases of Vibrio vulnificus for the year. However, following the storm, the county reported 14 confirmed cases, a large increase due to flooding transporting the bacteria into regions where people were cleaning up or returning to their homes.

A similar pattern emerged in other Tampa Bay-area counties, such as Hillsborough, where the number of cases increased from one before the storm to eight thereafter. Health officials gave many cautions to locals to avoid flooding and be wary of contaminated seafood, but infections increased in the weeks following the hurricane.

Floodwaters: Mixing Fresh and Saltwater and its Effects

The mixing of fresh and saltwater during a cyclone is a major element in the spread of Vibrio bacteria. Normally, Vibrio vulnificus thrives in coastal saltwater habitats, but when freshwater from rivers, storm drains, or rain runoff interacts with seawater, it produces brackish water, which Vibrio bacteria prefer even more.

Brackish water, with its intermediate salinity, is an ideal environment for bacteria to multiply rapidly. This water also contains other dangerous compounds, such as sewage and chemicals, which pollute the environment and pose a variety of health problems in addition to Vibrio infections.

Floodwaters from high rainfall or storm surges serve as a carrier for Vibrio bacteria. They disperse the germs to areas where they would not ordinarily be present, such as inland regions or areas that do not come into contact with coastal waters. Many individuals believe that floodwaters are merely rains and do not understand they include deadly bacteria until it is too late. As a result, workers cleaning up after a storm may come into close contact with contaminated water, increasing their risk of infection.

In addition to direct contact with floodwaters, there is fear over the contamination of local fish supplies. Flooded waters with Vibrio bacteria can harm marine life, especially shellfish such as oysters, which filter water as they feed. When these mussels are gathered from hazardous waterways, they can transport bacteria into homes and restaurants, endangering anyone who consumes them raw or undercooked. This is why health advisories issued after hurricanes frequently include warnings about consuming local seafood, particularly in the immediate aftermath of a storm when water quality testing may not have been completed.

It's important to realize that not all floodwaters are dirty or muddy, which might create a false sense of security. Vibrio bacteria are minute, and even water that appears clean or clear can contain a high bacterial load. To limit the danger of infection, it is vital to take precautions like as wearing protective garments and waterproof bandages, as well as avoiding direct contact with floods wherever feasible.

Hurricanes and flooding can have a significant impact on the spread of Vibrio vulnificus infections. The mixture of salt and freshwater, combined with warm temperatures, provides an ideal growing habitat for these hazardous bacteria. The aftermath of storms Ian and Helene highlight the substantial public health dangers posed by Vibrio infections, particularly when people consume polluted water or seafood. Staying informed, heeding health warnings, and adopting practical safeguards are all important factors in lowering the risk of these potentially fatal illnesses during natural disasters.

Identification of Vibrio Infections

Recognizing a Vibrio vulnificus infection early is critical because the bacteria can spread swiftly and cause serious, perhaps fatal, problems. Understanding the signs and knowing when to seek medical assistance might make the difference between a complete recovery and major health complications, such as surgery or amputation.

Vibrio infections can affect several sections of the body, and symptoms range based on the type of infection. This chapter will go over the common symptoms of Vibrio vulnificus infections, how to differentiate them from other bacterial diseases, and when to seek medical attention.

Symptoms of Vibrio vulnificus Infections

1. *Gastrointestinal symptoms*

Vibrio vulnificus is usually linked to gastrointestinal problems after consuming raw or undercooked shellfish, particularly oysters. These symptoms typically appear within 24 to 48 hours of ingesting contaminated food. The most prevalent signs are:

- <u>Diarrhea:</u> This is frequently the first and most prominent symptom. If not treated properly, the diarrhea might become watery and persistent, leading to dehydration.

- <u>Abdominal cramping:</u> This can be minor or severe, and it commonly resembles food sickness or a stomach bug.

- <u>Nausea and vomiting:</u> These symptoms, together with diarrhea, can deplete the body's fluids, causing discomfort and weakness.

- Fever: As the body fights the infection, it may develop a low-grade fever.

These gastrointestinal symptoms are often self-limiting, meaning they will disappear on their own in healthy people. People with compromised immune systems, liver illness, or other chronic disorders may experience more severe symptoms, such as dehydration or a spreading infection, which can lead to problems.

2. *Symptoms of a Bloodstream Infection*
In some situations, Vibrio vulnificus can enter the bloodstream, causing septicemia, which can be fatal. People with compromised immune systems or pre-existing health issues are more likely to get Vibrio bloodstream infections. Symptoms can appear quickly, frequently

within hours after exposure to the pathogen, and include:

- <u>Fever and chills:</u> High fever is common and can appear quickly, typically accompanied by severe chills and shaking.

- <u>Low blood pressure:</u> As the infection progresses, the bacteria can cause a severe reduction in blood pressure, leading to shock, a potentially fatal condition in which important organs do not receive enough oxygen.

- <u>Blistering skin lesions:</u> These lesions may develop as the infection progresses through the bloodstream. The skin may develop painful, fluid-filled blisters, which are commonly found on the legs, arms, and torso.

- <u>Confusion or dizziness:</u> Low blood pressure and temperature can cause confusion, dizziness, or disorientation, all of which are symptoms of a serious infection of the brain and other organs.

A bloodstream infection caused by Vibrio vulnificus necessitates rapid medical attention, which frequently includes antibiotics administered via an IV. Delayed treatment might lead to serious consequences or death, particularly in fragile patients.

3. *Signs of Wound Infection*

Wound infections are one of the most severe and visible types of Vibrio vulnificus infections. These usually happen when open wounds, cuts, or minor scratches are exposed to contaminated water. Wound infections can spread quickly, resulting in a condition known as necrotizing fasciitis, in which the tissue surrounding the wound dies. Early signs of wound infection include:

- <u>Redness and swelling surrounding the wound:</u> This is frequently the first indication of infection. The region surrounding the wound may turn red, bloated, and heated to the touch.

- <u>Discomfort that worsens over time</u>: The discomfort may begin mildly, but as the infection progresses, it can become acute and throbbing.

- <u>Rapid tissue breakdown</u>: As the infection worsens, the skin and tissue surrounding the incision may begin to disintegrate, turning black or producing ulcers.

- <u>Discharge or fluid leakage:</u> Infected wounds frequently flow pus or other fluids, which can have an unpleasant odor.

- <u>Fever</u>: As the body fights the illness, a fever may develop, indicating that it is spreading.

If left untreated, Vibrio wound infections can quickly worsen, necessitating surgical intervention to remove dead or contaminated tissue. In the most severe cases, amputation may be required to keep the infection from spreading to other regions of the body.

Identifying Vibrio Infections from Other Bacterial Infections

It is critical to distinguish a Vibrio vulnificus infection from other common bacterial illnesses, especially since symptoms such as redness, swelling, and fever are not exclusive to Vibrio. For example, Staphylococcus and Streptococcus bacteria can cause skin infections, but they progress more slowly than Vibrio.

One significant distinction between Vibrio infections is the rapid onset and intense tissue destruction, which can occur within hours to a day of exposure. If someone has recently been into contact with warm coastal waters or consumed raw seafood and develops any of the symptoms listed above, particularly after exposure to floodwaters following a hurricane, Vibrio should be considered a possible cause.

Doctors commonly diagnose Vibrio infections by collecting a sample of the infected site, blood, or stool to confirm the presence of the bacteria. This is important since the treatment for Vibrio may differ from that for other bacterial illnesses. Early detection is critical to avoiding problems and ensuring the appropriate medications are provided.

When To Seek Medical Attention

Timing is essential when dealing with Vibrio vulnificus infections. Even minor symptoms can soon deteriorate, and waiting too long to seek medical assistance might result in significant problems. It is critical to seek medical attention promptly if:

- You get severe diarrhea, nausea, or vomiting after consuming raw or undercooked fish.

- You observe redness, swelling, or increased pain near a wound, particularly if it has been exposed to seawater or flooding.

- Blisters or sores appear on the skin, especially if they are quickly deteriorating or accompanied by fever and chills.

- You feel weak, dizzy, or confused after experiencing any of the symptoms listed above, as they could be signals of sepsis or a spreading infection.

People with chronic illnesses such as diabetes, liver disease, or immunological diseases should be very cautious. These people are more likely to develop serious consequences from Vibrio vulnificus, therefore they should seek medical attention as soon as symptoms appear.

Diagnosing Vibrio Infection

Because Vibrio vulnificus infections can be fatal, diagnosing them accurately and quickly is critical. The sooner it is recognized, the greater the prospects of controlling the symptoms and avoiding complications.

Medical Tests for Diagnosis

When a patient exhibits symptoms consistent with a Vibrio vulnificus infection, such as gastrointestinal difficulties, fever, or a rapidly deteriorating wound, clinicians must act swiftly to establish whether Vibrio bacteria are present. Several medical tests are used to confirm the diagnosis.

1. Blood Cultures:
If a patient is suspected of having a bloodstream infection, such as sepsis, doctors will draw blood to check for the presence of Vibrio vulnificus. The bacteria

can spread swiftly through the circulation, thus early discovery via blood cultures is crucial for therapy. This test reveals the exact bacteria causing the infection and aids in the selection of appropriate medications.

2. <u>Wound Culture:</u>

Doctors will obtain a sample of fluid or tissue from the diseased location in patients who develop a wound infection, particularly after exposure to seawater or brackish floodwaters. This sample is cultivated in a laboratory to determine the bacterial strain. Because Vibrio vulnificus can cause necrotizing fasciitis, a condition in which tissue dies quickly, wound testing should be performed as soon as possible to avoid further tissue damage and the need for amputation.

3. <u>Stool Sample:</u>

When Vibrio vulnificus is suspected after eating contaminated seafood, a stool sample is typically taken. Diarrhea and stomach cramps are frequent symptoms of a gastrointestinal infection. Testing the stool helps to establish the presence of Vibrio bacteria, distinguishing it from other foodborne infections such as Salmonella and E. coli.

4. <u>Imaging Tests:</u>
While imaging tests such as X-rays or MRIs are not typically required for diagnosis, they may be utilized if deeper tissue infections or gas in the tissues are suspected, as these can suggest necrotizing fasciitis. This allows doctors to examine the severity of the illness and decide whether surgical intervention is required to remove dead tissue.

These tests, while effective, must be correctly interpreted by medical specialists. This necessitates that healthcare providers use a patient's recent behaviors, such as swimming in coastal waters, hurricane flood exposure, or eating raw seafood, to inform their testing recommendations.

Importance of Early Detection

The sooner a Vibrio vulnificus infection is detected, the greater the patient's chances of recovery. Early discovery enables doctors to begin antibiotic treatment immediately, which is the primary defense against the bacterium. Delays in diagnosis, particularly for wound or bloodstream infections, can have serious repercussions, such as limb amputation or death.

1. Preventing Sepsis:

If Vibrio vulnificus enters the bloodstream, it can induce sepsis, a potentially fatal condition in which the body's response to infection damages tissue and causes organ failure. Early identification using blood cultures and timely antibiotic therapy can prevent sepsis from progressing and lower the risk of death.

2. Save Tissue and Prevent Amputation:

When Vibrio enters a wound, early detection is critical to preventing the infection from spreading. When necrotizing fasciitis develops, the germs can kill skin and muscle tissue within hours. Early detection enables doctors to treat with medications and, if necessary, surgical excision of contaminated tissue before more harsh measures such as amputation are required.

3. <u>Lowering Misdiagnosis Risks:</u>
Infections with Vibrio vulnificus might resemble other frequent illnesses including cellulitis or gastrointestinal food poisoning. Early, targeted diagnostic tests help identify Vibrio infections from other bacterial disorders, ensuring that patients receive appropriate therapy as soon as possible.

<u>Examples of Misdiagnoses</u>
Despite breakthroughs in testing, Vibrio vulnificus infections can and do go misdiagnosed. There have been multiple documented situations in which Vibrio infection symptoms were misdiagnosed as

other illnesses, resulting in treatment delays and serious sequelae.

1. <u>Instance of Mistaken Cellulitis:</u>
One patient, who acquired a red, swollen lesion after wading through floodwaters, was initially diagnosed with cellulitis, a common bacterial skin illness. The doctor ordered oral antibiotics, which are normally used to treat less serious illnesses.

The patient's condition worsened during the next 24 hours, and the wound deteriorated fast. After a reevaluation, it was revealed that Vibrio vulnificus was the reason, and the patient needed emergency surgery to remove dead tissue. Early Vibrio testing could have detected the infection sooner, saving the patient from unnecessary pain.

2. <u>Misdiagnosed as food poisoning:</u>
In another example, a man who ate raw oysters experienced severe gastrointestinal

pain and diarrhea. He was diagnosed with food poisoning and sent home with orders to relax and stay hydrated. However, his symptoms persisted, and two days later he was admitted to the hospital with sepsis.

By that point, the bacteria had already entered his bloodstream, necessitating a more complicated and lengthy therapy. A simple stool test on the initial visit may have identified the Vibrio bacterium, allowing for earlier management and a far faster recovery.

3. <u>Unable to Recognize Necrotizing Fasciitis:</u> A third case included a woman who contracted a nasty wound infection during a fishing expedition. She was originally treated for a routine bacterial illness, but her symptoms worsened quickly. The doctors failed to notice the symptoms of necrotizing fasciitis, which is characteristic of Vibrio wound infections.

When the right diagnosis was obtained, the infection had progressed extensively and she needed repeated surgeries to remove contaminated tissue. An earlier identification through wound culture, as well as a better understanding of her previous behaviors, could have avoided such grave effects.

These examples highlight the necessity of not just early detection, but also the involvement of healthcare providers in considering Vibrio vulnificus in patients who have been exposed to coastal or floodwaters or consumed raw seafood. Misdiagnosis is especially harmful with this infection because of its rapid development.

Finally, early and precise diagnosis of Vibrio vulnificus infections is critical for avoiding serious health consequences. Doctors can detect Vibrio infections early by performing blood, wound, and stool cultures and taking into account the patient's recent activity. Avoiding misdiagnosis can dramatically enhance recovery prospects by lowering the danger of sepsis, tissue death, and amputation. Early detection is critical for both patients and healthcare providers.

Treatment Options

When it comes to Vibrio vulnificus infections, prompt and efficient treatment is essential. The germs can swiftly spread throughout the body, inflicting significant harm, and the infection can be fatal if not treated promptly.

Overview of Treatment Methods

Once Vibrio vulnificus has been detected, the treatment plan will be determined by the type and severity of the illness. Most patients require a combination of medications to limit the bacteria's growth, and in more severe cases, surgery to remove diseased or dead tissue. In addition, supportive care may be required to treat symptoms and complications such as sepsis or organ failure.

1. <u>Antibiotics:</u>
Antibiotics are the initial line of protection against Vibrio vulnificus. They function by destroying bacteria and preventing them from growing, allowing the body to heal. Antibiotics are prescribed based on the type of infection—gastrointestinal, bloodstream, or wound. Doctors typically prescribe a mix of medicines to provide better protection against the bacteria's rapid development.

2. <u>Surgical intervention:</u>
In some cases, particularly those involving necrotizing fasciitis (flesh-eating disease) or severe wound infections, antibiotics alone may be insufficient. Surgery may be required to remove infected tissue, especially if the infection has caused significant tissue damage. This technique can be lifesaving because it keeps the infection from spreading to other parts of the body. Early surgery can significantly improve a patient's recovery.

3. Supportive care:

Patients with serious infections may also need supportive care. This includes treating symptoms such as fever, dehydration from diarrhea, and pain. In cases of bloodstream infections or sepsis, further measures such as intravenous fluids, blood pressure support, and organ function monitoring may be required.

Antibiotics and their effectiveness

Antibiotics are quite successful in treating Vibrio vulnificus if administered early in the infection. However, because the bacteria proliferate and spread rapidly, delaying therapy drastically reduces their effectiveness. In extreme situations, doctors generally administer antibiotics intravenously (via an IV) to ensure they reach the bloodstream.

1. <u>Commonly Used Antibiotics:</u>
The two most widely used antibiotics for Vibrio vulnificus are doxycycline and a third-generation cephalosporin such as ceftriaxone. These antibiotics are powerful against bacteria and assist in keeping the infection under control. Some cases may necessitate a combination of antibiotics to completely clear the germs, especially if there is concern about resistance or the illness has spread to different places of the body.

2. <u>Challenges with Antibiotic Resistance:</u>
While Vibrio vulnificus is often responsive to medications, there is rising worry about antibiotic resistance. Resistance complicates treatment since conventional medications may not be as effective, necessitating the use of stronger or alternative antibiotics. This is one of the reasons why early detection and antibiotic treatment are critical—they prevent bacteria from becoming more resistant.

3. <u>Time of Treatment:</u>
The timing of antibiotic therapy is critical. Most individuals respond favorably to antibiotics when treated within the first 24 hours of infection. However, after the infection has moved into the circulation or begun to cause tissue death (as in necrotizing fasciitis), antibiotics may be insufficient, necessitating additional treatment.

When Surgical Intervention is Necessary

In certain severe situations, surgery becomes an essential component of the therapeutic regimen. This is especially true for wound infections that proceed to necrotizing fasciitis, a condition in which the tissue surrounding the wound dies quickly. Antibiotics can prevent bacteria from multiplying, but they cannot cure existing tissue damage. Surgery surgically removes dead or contaminated tissue, preventing the illness from spreading to healthy sections of the body.

1. <u>Necrotizing fasciitis:</u>
Necrotizing fasciitis is the most serious consequence of a Vibrio vulnificus wound infection. It frequently needs rigorous surgical debridement, which involves removing dead or infected tissue to keep the germs from spreading deeper into muscles and other tissues. This is frequently done in phases as doctors determine how much tissue has been impacted. If too much tissue is removed, patients may require skin grafts or reconstructive surgery once the infection has been treated.

2. <u>Amputation:</u>
In severe cases where the illness has progressed too far, surgeons may have to amputate the infected limb to save the patient's life. This is uncommon, but it can occur if the illness is not treated promptly or

spreads rapidly despite medical intervention. The decision to amputate is always agonizing, but it can keep bacteria from spreading to crucial organs or the bloodstream, where they could be lethal.

3. Post-Surgical Recovery:
Recovery from surgery is determined by the severity of the infection and the amount of tissue removed. Patients who have surgery for necrotizing fasciitis or other serious infections may need long-term rehabilitation, including physical therapy, to regain function and mobility in the affected area. If nerve injury develops, some people may lose their ability to move or feel their affected limb permanently.

Long-Term Consequences of Severe Infection

Survivors of severe Vibrio vulnificus infections may not be able to resume regular life immediately. The infection's long-term effects can vary, but they are typically severe. Patients who get necrotizing fasciitis or lose a limb through amputation may require months or even years of rehabilitation. In some circumstances, the psychological impact can be equally significant as the physical one.

1. Physical recovery:
Physical recovery from a severe Vibrio infection is slow, especially for individuals who need surgery. Patients may experience prolonged pain, scarring, and reduced movement. In circumstances where considerable amounts of tissue were removed, patients may require reconstructive surgery or skin transplants. Prosthetics may be required after amputation, and some patients may struggle to adjust to life without a limb.

2. Mental and emotional impact:
Surviving a serious infection such as Vibrio vulnificus can have long-term emotional

consequences. Anxiety, sadness, and post-traumatic stress disorder (PTSD) can result from the trauma of dealing with a life-threatening illness, the agony of surgery, and the loss of limb function. Patients are frequently urged to seek counseling and support groups to help them process their experiences and adjust to any physical changes.

3. <u>Recurrent Infection:</u>
Recurrent Vibrio infections are possible, however rare, particularly in people with impaired immune systems or chronic liver illness. Those who reside in coastal locations or are frequently exposed to water settings should be cautious about wound care and avoid conditions that could lead to re-infection.

Vibrio vulnificus infections require a multifaceted therapy that includes antibiotics, surgery, and supportive care. While many patients recover with prompt treatment, the risks of waiting are substantial, and the consequences can be serious.

Prevention Strategies

Preventing Vibrio vulnificus infections is critical, especially for people who are at a higher risk of serious sequelae. While the bacteria exist naturally in marine habitats, recognizing the conditions that cause infections can dramatically lower the likelihood of developing one.

<u>Understanding the Risk Factors</u>

Vibrio vulnificus thrives in warm, brackish waters, which are commonly found along coastal locations. The bacteria are more abundant in warmer months (May to October) and in locations that have undergone flooding or heavy rain. While everyone can become infected, certain people are predisposed to more severe forms of the disease.

1. <u>Compromised Immune Systems:</u>
People with compromised immune systems, particularly those with chronic liver disease,

diabetes, or cancer, are at a higher risk. This is because their body's ability to fight infections has weakened. Those with liver issues are especially vulnerable since the liver helps filter toxins from the body, and decreased liver function can allow the germs to spread more quickly.

2. <u>Open wounds or skin lesions:</u>
Vibrio vulnificus can enter the body via tiny skin breaks like cuts, scrapes, or insect bites. People with open wounds who come into contact with contaminated water are highly susceptible to infection. Even if the cut appears minor, exposure to seawater or floodwater where Vibrio grows could have catastrophic repercussions.

3. <u>Raw or undercooked seafood consumption:</u>
Another important concern is consuming raw or undercooked seafood, particularly oysters, which can contain high levels of Vibrio bacteria. People with immune system

disorders should avoid raw fish to avoid sickness.

Avoiding Exposure to Contaminated Water

Reducing exposure to potentially polluted water sources is crucial for preventing Vibrio infections. This is especially critical during natural catastrophes such as hurricanes, which can cause bacterial blooms in water due to the mixing of fresh and saltwater, as well as the entrance of pollutants.

1. <u>Avoiding Flood Waters:</u>

Following strong rains or hurricanes, health experts frequently warn against wading across floodwaters. These waters may be contaminated not just with Vibrio, but also with other hazardous bacteria, sewage, or debris. It is critical to follow these precautions and avoid direct contact with floodwater, especially if you have open wounds or damaged skin. When navigating flood zones, wear protective apparel such as waterproof boots and gloves to help reduce contact.

2. <u>Swimming in Safe Waters</u>

During the summer, when Vibrio bacteria are more widespread, it is critical to be cautious about where you swim. Public health authorities monitor water quality and frequently issue cautions if bacteria levels are very high. Avoid swimming in locations that have recently been flooded or where the water quality is poor. If you go swimming, make sure to properly wash any exposed skin afterward, especially if you have any wounds or scratches.

3. <u>Immediate Wound Care:</u>

If you have a wound and want to visit coastal waters or locations prone to Vibrio, cover it with waterproof bandages before going. If your wound comes into contact with water, you should clean it completely with soap and clean water as soon as possible. Keeping wounds clean and dry might help prevent bacteria from entering your body.

Safe Seafood Practices

Seafood, particularly shellfish such as oysters, is a common vector of Vibrio bacteria. While seafood is generally safe to consume when properly prepared, eating raw or undercooked fish raises the risk of Vibrio infections. Adopting safe seafood practices can greatly reduce the risk of infection, especially for people with underlying medical conditions.

1. Cook Shellfish Thoroughly

Heat kills Vibrio bacteria, therefore fully boiling shellfish is the most effective strategy to prevent infection. Boiling oysters, clams, and mussels for at least three minutes, or frying, grilling, or baking them to an adequate internal temperature, will kill hazardous germs. If you prefer eating oysters on the half shell, consider switching to fully cooked alternatives, especially if you are at a higher risk.

2. <u>Handle Seafood Carefully:</u>
Seafood should be handled and stored properly. Keep raw seafood cool (below 40°F) until ready to cook, and keep raw and cooked foods separate to prevent cross-contamination. To limit the danger of bacterial contamination, properly wash your hands, utensils, and surfaces after handling raw seafood.

3. <u>Check for public health advisories:</u>
Before ingesting seafood, especially in coastal locations, consult your local public health advisory for contamination levels. Shellfish harvesting locations are frequently examined for bacterial levels, and closures or warnings may be issued if conditions are deemed unhealthy. Following these

guidelines can help you avoid unintentional exposure to tainted seafood.

Educating Vulnerable Populations

While Vibrio vulnificus can damage everyone, particular populations are more likely to develop serious illnesses. Educating these groups about the hazards and preventive measures is critical to lowering infection rates and providing timely medical care when needed.

1. <u>Persons With Chronic Conditions:</u>
Individuals with liver illness, diabetes, cancer, or other immune-compromising disorders must be especially cautious. For some people, even minor Vibrio infections can soon become fatal. It is critical to educate children on the dangers of raw seafood intake, the necessity of wound care, and the risks connected with water exposure.

2. The Elderly Population:
The elderly are especially prone to dangerous illnesses because their immune systems gradually decline as they age. Outreach activities that educate people on proper seafood intake, personal hygiene, and wound care can help lower their risk. Furthermore, urging children to seek medical attention at the first indication of infection may save lives.

3. Coastal Community and Workers:
People who live close to or work on water, such as fishermen or lifeguards, are more vulnerable to exposure. These people should be warned about Vibrio dangers on a frequent basis, particularly following storms or flooding. Regular health checks, as well as access to first aid materials to treat wounds immediately, can help prevent dangerous infections.

4. <u>Tourism and Visitors:</u>
Vacationers in coastal locations may be unaware of the risks posed by Vibrio bacteria. Hotels, restaurants, and public health departments can all contribute by offering information on safe seafood practices and warning guests about swimming and water safety, particularly after a storm.

Preventing Vibrio vulnificus infections necessitates a mix of awareness, practical precautions, and responsible conduct, especially among those at higher risk.

The Role of Public Health

Public health systems play an important role in illness control and management, including Vibrio vulnificus. These organizations serve as the first line of defense against large-scale outbreaks, tracking cases, and informing the public. Their job is especially vital during natural catastrophes and in coastal locations where Vibrio bacteria can grow.

How Health Departments Track and Report Vibrio Cases

Monitoring and tracking infections such as Vibrio vulnificus necessitates collaboration across municipal, state, and national health authorities. This surveillance is critical because prompt reporting can help prevent the bacteria from inflicting extensive damage, particularly in conditions that enhance the risk of infection, such as warm weather and post-hurricane floods.

1. Surveillance and case tracking

Health departments use surveillance methods to track Vibrio cases. They collaborate closely with hospitals, clinics, and laboratories to collect information on suspected and confirmed cases. When a patient is diagnosed with a Vibrio infection, healthcare professionals must notify their local health department, which then communicates the information with state and national authorities like the Centers for Disease Control and Prevention (CDC). This enables health professionals to follow infection patterns and outbreaks in real time, allowing them to offer timely notifications and advisories.

2. Environmental monitoring

In addition to tracking human cases, health agencies monitor environmental conditions that may contribute to the spread of the Vibrio bacterium. Water quality testing is a critical component of this effort. Public health officials analyze coastal waters on a regular basis, particularly during the warmer months, to detect increased amounts of bacteria. When levels become too high, swimming advisories or shellfish harvesting bans may be issued to protect the public from exposure.

3. Data Reporting & Analysis
Public health agencies collect and analyze data to spot trends. This research can detect patterns, such as increased infection rates in specific places or at certain times of the year. Such insights help drive public health measures, allowing officials to direct resources where they are most needed. For example, if a rise in infections is found during a hurricane, health agencies can respond by issuing targeted warnings about avoiding flooding and raw seafood intake.

Significance of Public Awareness Campaigns

One of the most successful methods for combating Vibrio infections is to educate the population. Even though Vibrio vulnificus is a naturally occurring bacteria, many infections can be avoided by taking basic precautions such as avoiding contact with contaminated water and properly cooking seafood. As a result, public health authorities place a high priority on awareness efforts to educate and safeguard the public.

1. Health Alerts & Advisories

Health authorities routinely issue advisories regarding potential Vibrio dangers, particularly following events such as hurricanes or during peak summer months when water temperatures increase. These advisories may include cautions about floodwaters, no-swim zones, and seafood safety requirements. The goal is to ensure that both residents and visitors are aware of the hazards and take appropriate safeguards.

2. Educational Material and Outreach

Effective public health campaigns use a range of platforms to reach different groups. Educational materials, including as booklets, posters, and social media content, are intended to deliver important information about Vibrio straightforwardly and understandably. Public health agencies frequently work with local organizations, schools, and community centers to distribute these materials, ensuring that vulnerable populations, such as coastal residents and immunocompromised people, have access to vital information.

3. Engaging Vulnerable Populations

Certain groups, such as the elderly or those with compromised immune systems, are more likely to develop serious Vibrio infections. Public health initiatives frequently target these susceptible communities, offering specific advice on how to be safe. This might include explicit cautions about avoiding raw seafood or tips on how to care for wounds if exposure to possibly contaminated water is unavoidable.

4. <u>Public Education Following Natural Disasters</u>

Hurricanes and other natural disasters frequently generate conditions that allow Vibrio bacteria to grow. Public health campaigns become especially crucial during these occurrences since individuals may be exposed to flooding or must rely on local fish sources.

Following a hurricane, health agencies may conduct targeted initiatives to educate the public about the heightened risk of infection. These campaigns may employ radio, television, or digital media to reach a huge number of people quickly.

Community Resources for Education and Assistance

In addition to awareness programs, public health agencies and other groups offer a variety of community tools that educate and support people who are at risk of Vibrio infections. These resources ensure that the public has access to credible information, preventive methods, and emergency aid in the event of an infection.

1. Hotlines & Websites

Many public health agencies maintain hotlines or provide thorough information on their websites, including updates on local water quality, food safety advisories, and infection prevention measures. Individuals can use these platforms to ask questions and receive professional advice on everything from wound care to seafood eating, allowing them to make more educated decisions during times of high risk.

2. Community Health Centres and Outreach Programmes

Local health facilities frequently play an important role in educating the public about Vibrio vulnificus, particularly in coastal or flood-prone areas. These centers may provide free or low-cost training sessions, workshops, and materials that explain how to spot the symptoms of Vibrio infections and what steps to take if someone believes they have been exposed. Outreach initiatives also assist persons who may not have easy access to standard healthcare services.

3. Partnering with Local Organizations

Public health agencies routinely collaborate with community organizations, schools, and local governments to broaden the impact of their campaigns. These collaborations may include putting up information booths at community events, arranging health fairs, or holding instructional sessions in schools. This concerted effort ensures that individuals have the information they need to protect themselves and others from Vibrio illnesses.

4. Assistance during outbreaks
During outbreaks, public health departments work together to control the issue. This involves advising healthcare providers on how to identify and treat infections, administering antibiotics as needed, and assisting hospitals and clinics in successfully responding to rising instances. In some cases, such as after a large hurricane, public health professionals may collaborate with national organizations such as the Federal Emergency Management Agency (FEMA) to provide resources and assistance to impacted areas.

Public health is critical in preventing and treating Vibrio vulnificus infections. Health departments help lower the risk of illnesses by combining monitoring, public awareness campaigns, and community outreach. They also give vital education to vulnerable populations.

Individuals and communities can better protect themselves from this harmful bacterium by understanding the efforts done at the public health level, particularly in times of increased risk, such as after natural disasters or during the warmer months. Awareness, education, and prompt action are critical to reducing Vibrio's impact on public health.

What to Do After a Flood

Flooding can leave behind a hazardous mixture of chemicals, including deadly bacteria such as Vibrio vulnificus. Following a flood, you must take certain care to protect yourself and your family from dangerous diseases. Knowing what to do when you return home, how to assess the level of contamination, and how to clean and sterilize impacted areas can help to reduce health hazards.

Steps to Follow When Returning Home After Flooding

Returning home after a flood can be stressful, but safety comes first. Floodwaters can contain a range of risks, including chemicals, sewage, and pathogens, so it's critical to take a methodical approach while examining your home.

1. Wait for official clearance

Before returning to your property, ensure that local authorities have declared the area safe for return. Even after floodwaters recede, invisible threats may linger, such as weakened infrastructure or standing water tainted with bacteria like Vibrio vulnificus. For information on when it is safe to return, check local news or follow health and emergency services advice.

2. Wear protective gear

Always wear protective gear when you come home. This comprises waterproof boots, gloves, long sleeves, and pants to reduce skin contact with contaminated water or surfaces. Even after the water has been removed, residual contamination can be found on surfaces such as walls, furniture, and appliances. If you have open wounds or scratches on your body, cover them and prevent exposing them to possibly contaminated regions.

3. Assess the structural damage

Before entering the house, check for structural damage. Flooding can undermine a building's foundation, walls, and roof, rendering it unsafe. Do not enter if you see cracks, sagging, or other signs that the house may collapse. If in doubt, check with a specialist to determine whether it is safe to enter the building.

4. Ventilate the area

Once inside, open the windows and doors to let fresh air flow. This helps to limit the buildup of toxic vapors from chemicals or mold, as well as the concentration of harmful germs. Proper ventilation also makes it easier to operate within the structure, especially when using cleaning chemicals during the sanitation process.

5. Document the damage

Before you begin cleaning up, take photos or videos of the damage to your home and valuables for insurance purposes. This is vital for insurance claims and may be valuable if health concerns occur in the future, as it provides a record of the conditions you experienced.

How to Assess Potential Contamination

Floodwaters include a variety of toxins that can be dangerous to humans. These include sewage, industrial waste, and germs like Vibrio vulnificus. Knowing how to properly inspect your home for potential contamination can help you avoid significant health risks.

1. Check for water damage

Any surface that has been in contact with floodwater should be considered polluted. This covers the floors, walls, furniture, and appliances. Contaminants may remain present long after the water has dried, so exercise caution when handling these goods. If in doubt, err on the side of caution and presume that surfaces are dangerous until thoroughly cleansed.

2. Look for mold growth

Floods frequently result in mold growth, which can begin within 24 to 48 hours of water contact. Mold can cause respiratory issues and aggravate allergies or asthma. Check for mold in areas where moisture has accumulated, particularly behind walls, under flooring, and in insulation. Visible indicators of mold growth include black, green, or gray patches on walls or other surfaces, as well as a musty stench. Even if mold is not visible right immediately, it can grow in hidden locations such as wall cavities, therefore a thorough check is required.

3. Inspect for biological contamination

If floodwaters flooded your home, they most certainly brought with them biological contaminants such as sewage, dead animals, and bacteria like Vibrio vulnificus. These offer substantial dangers, especially if you come into contact with them through open wounds or ingestion. If you think that sewage or other hazardous items have entered your house, contact professionals for assistance with the cleanup.

4. <u>Testing Water Sources</u>

Flooding can endanger drinking water supplies. If your home's water originates from a well or if public water systems were compromised, you must test the water before consuming, cooking, or cleaning. While boiling water can help eliminate some bacteria, chemicals, and other contaminants requires more thorough treatment. Until tests indicate that your water supply is safe, it is best to drink bottled water or water from a trusted source.

Tips for Cleaning and Sanitizing Affected Areas

Cleaning and sanitizing your home after a flood is an important step towards ensuring safety. While it may be tempting to jump right into the cleanup, it is critical to follow the proper measures to minimize the spread of bacteria and lower the risk of diseases, particularly those caused by Vibrio vulnificus.

1. Remove Contaminated Item

Any porous materials, such as carpets, mattresses, and upholstered furniture, that have been saturated by floodwaters should be discarded. These materials can absorb bacteria and chemicals and are frequently difficult to sanitize completely. Non-porous materials, such as metal, glass, and hard plastics, may usually be cleaned and disinfected.

2. Clean with soap and water first
Before applying disinfectants, clean surfaces with soap and water. Cleaning eliminates dirt, mud, and debris, which can impair disinfectant effectiveness if not addressed first. Use a stiff brush to clean hard surfaces such as floors, walls, and countertops, then thoroughly rinse.

3. Disinfect Surfaces
After cleaning, disinfect all surfaces with a bleach-and-water solution (1 cup bleach for every 5 gallons). Pay close attention to areas that were in contact with floodwaters, including kitchens, bathrooms, and other high-traffic areas. Make sure to disinfect non-porous things such as toys, tools, and kitchen utensils. To ensure that bacteria are killed, allow the disinfectant to stay on the surface for at least 10 minutes before rinsing.

4. Completely dry the area
Mold and germs flourish in moist settings, so dry out your property as soon as possible. Use fans, dehumidifiers, and open windows to accelerate the drying process. If you suspect hidden moisture in walls or flooring, call a specialist to avoid long-term problems such as mold growth.

5. Handle Hazardous Materials Carefully
If dangerous materials, such as chemicals or raw sewage, have entered your home, you should hire a professional to remove them. To avoid injury, these items must be handled with great care. You should not attempt to clean them yourself unless you have the necessary protective equipment and skills.

6. Look for signs of infection
Even after a thorough cleaning, be cautious about your health. If you or anybody in your household experiences symptoms such as fever, chills, nausea, or evidence of infection

near cuts or wounds, seek medical assistance right once. It is especially crucial to monitor wounds that have come into touch with floodwaters, as Vibrio vulnificus infections can spread quickly.

Taking the appropriate precautions after a flood can protect you and your family from health concerns, including serious bacterial infections such as Vibrio vulnificus. You can dramatically reduce the risk of infection by following safety protocols while returning home, carefully assessing contamination, and implementing proper cleaning and sanitizing practices. Flood cleanup is about more than just repairing your property; it's also about protecting your health at a period of increased risk.

Personal Stories and Case Studies

When dealing with a potentially fatal virus such as Vibrio vulnificus, human stories can be quite effective. They provide real-world insights into the human experience of dealing with bacteria, emphasizing the challenges, accomplishments, and lessons learned along the journey.

Testimonials of People Affected by Vibrio Infections

For many persons who have contracted Vibrio vulnificus, the ordeal starts innocently. A tiny cut when wading in coastal waters, an overlooked scratch, or even consuming contaminated seafood might have serious repercussions. Hearing their tales is critical to understanding how quickly Vibrio illnesses can spread.

John's Experience: A Life-Changing Cut

John, a 52-year-old fisherman from Florida, had no idea a little injury on his leg would change his life. He had spent years working in the Gulf of Mexico and was familiar with its waters. While cleaning his fishing boat after a lengthy voyage, he scraped his leg against a sharp piece of metal.

At first glance, it appeared to be a regular wound—one he'd suffered previously. However, after 24 hours, his leg began to swell and turn red. He experienced a severe burning sensation, followed by fever and chills.

By the time John sought medical attention, the infection had progressed rapidly. Doctors identified him with a Vibrio vulnificus infection and immediately prescribed medication. Unfortunately, the bacteria had spread too far, and he had his leg amputated. John tells his tale to warn others: "I thought it was just a scratch. If I had known how dangerous the water could be, I would've never ignored it."

His example demonstrates how seemingly minor injuries can have terrible consequences if exposed to Vibrio-contaminated water. Quick action and awareness are critical to avoiding such serious results.

Maria's Experience: A Seafood Lovers' Nightmare

Maria, a 37-year-old mother of two, adored seafood and frequently ate oysters with her family. One summer evening, she and her husband decided to celebrate their anniversary at a local seafood restaurant noted for its fresh catch.

Maria became quite unwell within 36 hours of eating the oysters. She started vomiting, had diarrhea, and her entire body felt weak. Her skin also formed blister-like sores. Her husband hurried her to the emergency room, where she was diagnosed with a Vibrio vulnificus infection caused by raw oysters.

Maria's story highlights the dangers of eating raw or undercooked fish, particularly for those with underlying health concerns. Maria was unaware that her modest liver condition increased her chance of severe complications from Vibrio infections. "I didn't know I was in danger. Now, I only eat cooked seafood. It's not worth the risk," Maria told me.

These personal experiences are not only about catastrophes but also about learning and adapting actions to avoid future disasters. Both John and Maria emphasize the significance of recognizing symptoms early on and acting immediately.

Perspectives from Healthcare Professionals

Healthcare workers who treat Vibrio infections on a daily basis gain significant insights into the medical and practical aspects of these illnesses. Their stories emphasize the need for early detection and the specific challenges of treating serious bacterial illnesses.

Dr. Kelly's Experience: The Value of Quick Diagnosis

Dr. Kelly, an infectious disease expert in Louisiana, has treated multiple Vibrio vulnificus cases, especially after hurricanes and during warm months when the bacteria thrives in coastal waters. "The biggest challenge we face is the speed at which this bacteria can spread," she tells me. "Patients often delay seeking care because they think their symptoms are just a minor infection or food poisoning. By the time they come in, the infection has taken hold, and we're in a race against time."

Dr. Kelly emphasizes the necessity of public knowledge. She describes an instance in which a patient delayed too long to seek treatment for what he thought was a minor infection after eating raw oysters. "We were able to save him, but he lost the use of his hand due to the necrosis caused by the bacteria. Vibrio vulnificus is aggressive, but with early diagnosis and the right treatment, we can prevent these severe outcomes."

Nurse Amy: Education and awareness save lives

Nurse Amy, who works in a coastal clinic in Texas, has a similar perspective, emphasizing the value of education. "We need to educate people living in or visiting coastal areas, especially after floods or hurricanes," she explains. "Many of the patients we see aren't aware of the risks posed by floodwaters or contaminated seafood."

She demonstrates how simple precautions, such as wearing protective equipment and keeping wounds clean and covered, can help save lives. "You don't need to be paranoid, but you do need to be informed," she says. "Knowledge is your best protection. And for those with compromised immune systems, avoiding raw seafood and floodwaters is critical."

One of the most important lessons for both patients and healthcare providers is the significance of early detection and treatment. Acting swiftly can make the difference between life and death, especially with Vibrio vulnificus, which can cause immediate tissue damage and serious bloodstream infections.

Here are some major takeaways and tips from these experiences:

1. <u>Know the Risks:</u> People with underlying health issues, such as liver disease, diabetes, or compromised immune systems, are more susceptible to severe Vibrio infections. If you fall into these groups, avoid raw or undercooked seafood and exercise caution when near coastal waters, especially after a storm or flood.

2. <u>Identify the Symptoms:</u> Vibrio infections can manifest in a variety of ways, such as gastrointestinal symptoms, wound infections, or signs of sepsis. If you get symptoms such as fever, edema, redness, or blistering around a wound after being exposed to water or seafood, seek medical attention right once.

3. <u>Don't Delay Treatment:</u> Many people are hesitant to seek medical assistance because they believe their symptoms will resolve on their own. This delay may be dangerous. If you suspect a Vibrio infection, particularly if you have risk factors, seek medical

assistance right once. Early antibiotic therapy can help to avoid serious complications, such as amputation or death.

4. <u>Take Preventive Measures:</u> Do not expose open wounds to possibly contaminated water, especially in coastal locations or after flooding. If you must enter floodwaters, always wear appropriate protective clothes and properly clean any wounds that come into touch with water. Those at high risk should avoid raw fish and ensure that any seafood they ingest is thoroughly prepared.

5. <u>Stay Informed:</u> Education and awareness are essential. Understanding the risks posed by germs such as Vibrio vulnificus and how to protect yourself will help you avoid infections.

These anecdotes and insights paint a more accurate picture of how serious Vibrio infections may be, but they also inspire hope. The worst outcomes can often be prevented with education, early action, and prevention.

Resources and Additional Reading

Understanding Vibrio vulnificus and its related infections is critical for raising awareness, preventing infections, and providing treatment. There are numerous organizations and services available to help people, families, and communities learn more about this bacteria and how to protect themselves.

Organizations and Websites with More Information

1. The Centers for Disease Control and Prevention (CDC)
- Site: [CDC Vibrio](https://www.cdc.gov/vibrio/index.html)
- The CDC provides extensive information about Vibrio bacteria, including

transmission, symptoms, and preventative techniques. Their services include outbreak data, seafood safety standards, and instructional materials geared toward both healthcare professionals and the general public.

2. The World Health Organization (WHO)
- Site: [WHO Vibrio](https://www.who.int/news-room/fact-sheets/detail/vibrio)
- The WHO offers a global perspective on bacterial illnesses, including Vibrio. Their efforts are focused on public health policies, monitoring, and prevention measures that can be used globally.

3. National Oceanic and Atmospheric Administration (NOAA)
- Site: [NOAA FishWatch](https://www.fishwatch.gov/)
- NOAA provides insights into seafood safety and sustainability. They include information on the health concerns

connected with various types of seafood, as well as guidelines for safe intake.

4. American Society for Microbiology (ASM).
- Website: [ASM Vibrio Information] (https://asm.org/Articles/2021/May/Vibrio-vulnificus-and-Vibrio-parahaemolyticus).
- ASM publishes publications and studies about microbial pathogens, including Vibrio vulnificus. Their offerings include research updates and educational materials designed for both professionals and the general audience.

5. Local health departments
- Visit your local or state health department's website for region-specific information on Vibrio infections, such as statistics, preventative suggestions, and local seafood and water safety rules.

Recommended Readings and Studies on Vibrio Infections

1. Joe C. T. Lee and colleagues' article "The Vibrio vulnificus Infection: A Review"
- This comprehensive review delves deeply into Vibrio vulnificus, including biology, epidemiology, and clinical symptoms. It is a fantastic resource for individuals interested in the scientific background of the infection.

2. "The Role of Climate Change in the Emergence of Vibrio Species" by Arthur A. H.L.E.F.H. Tan et al.
- This research paper looks at how climate change affects the distribution and pathogenicity of Vibrio species, particularly Vibrio vulnificus. It is critical for understanding the larger environmental issues influencing public health.

3. "Vibrio Infections in the United States: The Role of Environmental Factors" by Robert J. L. Brown

- This study examines the relationship between environmental factors like as temperature and salt and the prevalence of Vibrio infections. It underlines the need to monitor and regulate these elements to avoid outbreaks.

4. The FDA's "Safe Seafood: Guidelines for Consumers"
- This document describes safe seafood consumption habits, such as proper cooking temperatures and storage methods. It's a practical handbook that explains how to eat seafood while reducing health hazards.

5. "Emerging Infectious Diseases and Vibrio" by the National Institute of Health (NIH)
- This book offers insights on developing infectious diseases, with an emphasis on bacterial pathogens such as Vibrio. It discusses the difficulties caused by antibiotic resistance and the significance of public health interventions.

Contact Details for Local Health Departments

Finding local resources can be critical for emergency assistance or questions about Vibrio infections. Here's how to contact your local health departments:

1. <u>Finding Your Local Health Department:</u>
- To find your state's health department, go to the [CDC State Health Departments](https://www.cdc.gov/publichealthgateway/healthdirectories/healthdepartments.html) webpage.

2. <u>State Health Departments:</u>
- Each state has its health agency, which can provide information about Vibrio diseases, seafood safety, and local health advisories.

3. <u>Local health departments:</u>
- For specific questions, look up your county or city's health department. They can provide specific information and resources

for your area. Many municipal health agencies also provide hotlines for immediate concerns about infectious diseases.

Individuals and communities can learn more about Vibrio vulnificus and its hazards by using the resources and information offered in this chapter. Educating yourself and others on prevention, symptoms, and treatment can help save lives.

Staying informed through reliable organizations, recommended readings, and local health agencies enables consumers to make better decisions about seafood eating and water exposure. As awareness spreads, communities can collaborate to mitigate the effects of this hazardous pathogen.

Awareness and education are critical in combating the threat posed by Vibrio vulnificus. Communities must be aware of the illnesses that cause epidemics and the signs that require medical attention.

Individuals who are educated about safe seafood practices, the dangers of floods, and the significance of sanitation following potential exposure can better protect themselves and those around them.

Public health campaigns can have a substantial impact on behavior by emphasizing the importance of vigilance, particularly during hurricane season or following severe rainfall. These activities can help to promote a culture of awareness, in which people take proactive steps to reduce their risks.

Everyone plays a role in raising awareness about Vibrio vulnificus. Sharing information with friends and family can help them realize the hazards and actions they should take. Participate in community conversations or local events centered around water safety, seafood consumption, and health education. Consider volunteering with local health organizations dedicated to educating the public about infectious diseases.

We can contribute to a more informed society that values health and safety by sharing knowledge and promoting conversations. In an age where knowledge is more easily accessible than ever, making the effort to educate ourselves and others can have a tremendous impact on community health outcomes.

Awareness can save lives, and together, we can foster a culture that values education and preventive measures to combat illnesses like Vibrio vulnificus. Let us pledge to be attentive, knowledgeable, and ready to assist one another in protecting our communities from the hazards posed by this harmful disease.

www.ingramcontent.com/pod-product-compliance
Lightning Source LLC
Chambersburg PA
CBHW070249220526
45465CB00004B/1565